Tesla, Elon Musk, and the EV Revolution

An in-depth analysis of what's in store for the company, the man, and the industry by a value investor and newly-minted Tesla owner

Vitaliy Katsenelson

ISBN: 978-1-7358896-0-3

Published by Shabbos Goy Productions

REVIEWS FROM READERS

"I LOVE everything you write, but even benchmarked against your own lofty standards, this is a Magnum Opus that taught me more than I can adequately express!"
- Dan

"The way you think and the clarity with which you express your thoughts encourages the reader to think-and-debate along ... [this book] was devoured with delight!"
- Switesh

"Wow! I was impressed with your comprehensive analysis."
- Christine

"A fascinating read, especially with how the stock has performed since you wrote this."
- Joe

"I really enjoyed reading it. Not just the insights, but also the style. One night I planned to read the first chapter, but it sucked me in and I couldn't stop reading until I finished the whole analysis."
- Andras

"It is the most extensive reading on the topic that I've ever done... I found it to be quite comprehensive, thorough, and even handed."
- Harold

"I was thoroughly impressed with the depth of your research, you provide insights I would never (not even close) have come by on my own."
- Tony

"You did some heavy-lifting on your observations, analysis, & synthesis."
- Michael

"[Your book] opened my eyes to how much research is involved, not just the financials but the information on the companies activities."
- Jack

"An excellent, credible [work] backed up by what is clearly comprehensive research."
- Colin

DEDICATION

To all the Tesla optimists and sceptics out there

TABLE OF CONTENTS

Introduction

You don't really know the company until you buy the stock. It has happened to me a few times. We did hundreds of hours of research, bought a stock, and that act of buying activated new senses. I started seeing new angles. Something similar happened to me with Tesla, except I didn't buy the stock, I bought a car.

In late-June 2019, I bought a Tesla Model 3. Most people would just enjoy driving such a car. Not me – I wrote about it for two or three hours a day (usually early in the morning). During evenings and on weekends, I read anything I could find tangentially related to the electric vehicle (EV) industry. I discussed EVs with a lot of my friends, who helped me to shape and polish my thinking. Why did I spend so much time on this? I really wanted to understand the impact EVs will have on everything, from a drop in the demand for oil to an increase in the demand for other energy sources to the transformation of geopolitics to changes in the trucking industry and the implications for railroads to … the list goes on and will continue to grow.

There was also an element of natural curiosity. I learned a lot, and the process of learning brings incredible satisfaction.

There were also offensive and defensive business reasons. I am always looking for new investments, and a lot of opportunities will be created by the transition to EVs. On the other hand, EV disruption will claim its victims – stocks to avoid. Finally, I was able to develop several mental (thinking) models, which I will be able to apply to other investments.

I really hope you enjoy this.

- Vitaliy

Part 1: A Tsunami Is Coming at the Auto Industry

The test of a first-rate intelligence is the ability to hold two opposed ideas in mind at the same time and still retain the ability to function.

– F. Scott Fitzgerald

This book is going to raise more questions than it will answer. It will equally frustrate Tesla bulls and bears. I was thinking a lot about F. Scott Fitzgerald's quote about first-rate intelligence as I juggled opposing frameworks and conclusions.

My history with Tesla goes back to 2015. I put a deposit down on a Model 3 the day Model 3 was announced. I patiently waited for two years, but when I went to buy a car, my local Tesla store did not have any for me to test-drive, the four-wheel-drive version was not available, and the $35,000 car I was originally promised was suddenly being marketed by Tesla as a luxury BMW competitor and priced as such.

At that point, my $1,000 deposit looked like a two-year interest-free loan to Tesla, and I asked for my money back. As I later found out, I was not the only one who felt that way, as Tesla's 455,000-order backlog, which at the 2018 production rate should have lasted years, disappeared very quickly. Six months later, I was bombarded by emails from Tesla telling me that I didn't have to wait to buy a new car.
Fast-forward a year.

Tesla released the four-wheel-drive option, lowered and simplified its prices on the Model 3, and increased the driving range to 310 miles. By this time, my local Tesla store had plenty of Model 3s for me to test-drive. And drive I did. I had driven the Model S before, and I was pleasantly surprised to find that, despite not having air suspension, the Model 3 was even more pleasant to drive.

Finally, the price started to make sense. The four-wheel-drive version, with a 310 mile driving range and a premium interior, had a price out the door of $51,000; with $9,000 of federal and Colorado tax credits, the net cost to me was around $42,000. Yes, dear reader, you paid for almost a quarter of my car.

My car came with two keys that look just like credit cards. I have not used them at all. The car is connected to the Tesla app on my iPhone through the 4G network and Bluetooth. I can completely control the car from my iPhone app. If it is hot or cold outside, I can turn on the AC or heater from my phone. I can see where my car is at any time.

When I approach my Tesla, I don't have to do anything (my phone is in my pocket): The car magically opens. When I sit down, I don't have to push a start button; I just push the lever to D or R (drive or reverse) and start driving. Because no combustion is happening in the electric engine, my car makes no noise when I drive. When I am done, I put the lever on P (park), get out of the car, and it automatically locks.

Tesla made very radical interior design choices. Aside from buttons to open and close doors and windows, there are no physical buttons inside the car; all buttons and other controls are virtually placed on a giant tablet. I can control the stiffness of the steering wheel, turn the regenerative braking on and off, or set the quickness of acceleration as if choosing settings in a video game.

In fact, the car can also turn into a video game. The Model 3 comes with a dozen video games that you can play when the car is parked or waiting for a Supercharger to fill it up. The games are controlled by the real controls of the car. There is a driving video game that my daughters love where you use Tesla's real steering wheel and brake to drive a virtual car.

The Model 3 is really a computer, an iPad on wheels. The iPad-looking tablet and minimalistic design, which sits right in the middle of the front console, at first seemed a bit odd, but it felt natural once I was in the driver's seat.

The software interface of the tablet seemed very Apple-like; it was not designed by engineers for engineers (something Microsoft would do) but by human beings for human beings. The alarm clock radio controls of my wife's Honda Odyssey or my German SUV (now my wife's car) now feel a bit like dumb phones versus smartphones (an important theme of this article).

When you buy a traditional car, what you get on the day of purchase is what you get to use for the rest of the car's life. The Model 3, however, improves every month or so with software updates. These aren't just cosmetic user-interface changes to the display – after one of those updates, Model 3 owners woke up and discovered that their rear seats were heatable. The hardware was already there; the software update activated it. Tesla constantly releases new features, from arcade games to upgrades to the security system, through software updates.

I've owned the car for a bit over two months, and so far it is the best car I've ever had. Just as with a slight benefit of hindsight, it is easy to see how smartphones (whether or not made by Apple) are the future of mobile phones, I can see that electric cars are a tsunami that is coming straight toward the automotive and transportation industries and their supply chains; it's just hard to say yet when it will land onshore and wipeout (replace) the internal combustion engine (ICE) car. But it's clearly a question of when, not if.

Electric is the car of the future; that's as clear to me as rain.

Part 2: Why the Gasoline Car to the EV is Like the Horse to the Car

The complexity of the traditional ICE car is mind-boggling. There are several thousand moving parts that are interconnected by belts and gears and need to be constantly oiled and cooled. Air must be continuously pulled in to mix with gasoline to achieve combustion.

While the ICE engine is large, heavy, and "internally combusts" many times each second, the electric engine is exponentially simpler. It's much lighter and smaller, and you can count the number of moving parts on your right hand; it can be mounted right on the axle. This removes a lot of complexity in car design.

This is the punchline: The engine is the most complex and important part of the ICE car, and it is one of the least complex parts of the EV and, perhaps surprisingly, the least important one.

Let's pause for a second and admire the brilliance of electric car design. It took me owning the car for a few weeks to fully inhale its brilliance. An ICE engine needs to be revved up to 1,500–6,000 RPMs to get optimum torque, then shifting gears results in the desired speed. The electric engine instantly gets to maximum torque; by increasing or decreasing RPMs, the desired speed is achieved without any gears. None! Every time you add gears between the engine and the wheels, you add complexity and lose power. This is why ICE cars are only 30% efficient while EVs are 80% efficient.

Let's say you want to turn a two-wheel-drive ICE car into four-wheel drive. Now you need to transmit the power from the engine, which sits in the front of the car, to the back of the car. The power has to go through even more oiled metal parts to make it to the back wheels, so you lose power and reduce gas mileage.

The EV solution to four-wheel drive is very different: You just put another watermelon-sized engine on the rear axle and you double the power of the car.

This is where it gets interesting. The second engine adds some weight and cost, but it doesn't increase the complexity of the car and it results in better "fuel" efficiency. The second engine increases the efficiency and range of the vehicle because the EV actually generates electricity when you brake, through regenerative braking. Instead of pushing on the brake when you want to slow down, you let off the "gas" pedal. The car continues to roll by momentum and turns the electric engine, which now, instead of consuming electricity, generates electricity, recharges the battery, and also slows down the car. Thus, you get regenerative braking! And because two engines produce more electricity than one, the second engine increases the EV's range.

Because EV engines sit on top of axles, the whole concoction looks a lot like a skateboard with a battery lying flat between the axles. Tesla pioneered this design. It results in several benefits. First, the absence of a transmission means there is nothing bulging down the center of the car and thus there is more space inside the cabin.

Second, you have empty space up front, which can be used for storage – that's called the frunk. There is a significant safety benefit of the frunk: It serves as a giant bumper. In a front-end collision, the frunk absorbs the energy of the impact, whereas in an ICE car the engine absorbs the energy and then transfers it directly into the cabin, impacting the driver and front-seat passenger.

Third, the battery pack significantly lowers the Tesla's center of gravity, improving the driving experience and making the car less likely to flip.

Fourth, because electric engines produce greater initial torque, EV cars are quicker than ICE cars. My Model 3 goes from 0 to 60 in 4.4 seconds. Every time my wife gets into the passenger seat of my Model 3, she reminds me that I am a money manager, not a professional race car driver. The performance version that costs about $7,000 more than the version I bought does this in 3.2 seconds, putting it on a par with a super sports car like a Corvette.

Fifth, because electric engines work on electromagnetism and there is little friction, you don't need to oil them. There are very few moving parts, so electric cars require a lot less servicing than their ICE counterparts. Due to regenerative braking, even the brake pads should last much longer than on ICE cars.

I can keep going, but I'll stop with six: An electric car does not need a grille in the front because it doesn't have a radiator, as it doesn't need to cool down the engine, nor does it need to pump air into the engine. Therefore, the car is a lot more aerodynamic and thus again more efficient.

As a side note, it is too early to freak out about the impact electric cars will have on car dealerships and car repair and service businesses. The U.S. sells about 17 million cars a year, and there are about 180 million cars on the road today. If EV sales accounted for half (8.5 million units) starting today and continued at that rate for the next 10 years, at the end of that period the total EVs on the road would still be only half of all vehicles. Reality check: In 2018, only a few hundred thousand EV cars were sold. In the short run, EVs will have little impact on ICE servicing businesses; in the long run, those businesses will decline dramatically into a sunset.

Part 3: How EV Range Anxiety Will Give Way to an EV-Charging "Gold Rush"

Range anxiety is a real thing when you buy an EV or consider buying one, but charging has not been an issue for me at all. My Model 3 came with a portable charger. My garage, like most others, has a 110-volt outlet. At 110 volts, my Model 3 is charged at about four miles of range per hour. It would take three days to fully charge the car.

I hired an electrician, and $650 later my garage is wired with a 220-volt outlet (the same voltage your traditional washer and dryer use). If I bought a special charger from Tesla for $500, it would be hardwired into my 220-volt outlet and charge my car at 45 miles an hour. I use the portable charger that came with the car, plugged into the 220-volt outlet, and my car is charged at 30 miles per hour. It would take roughly 10 hours to fully charge the battery (remember, the total driving range is 310 miles). When I come home, I park my car in the garage and connect the charging cable, and an hour or two later the car is fully charged. I cannot tell you how much I do not miss gas stations.

Tesla recommends not fully charging on a daily basis. That's not good for the health of the battery, and regenerative braking does not kick in if the car is 100% charged. I drive 20 miles most days (the average American drives 30 miles per day). I charge my car to 270 miles on a daily basis, but I charged to 300 miles when my kids and I took our (almost) annual 360-mile road trip to see the Santa Fe Opera.

Tesla has about 1,600 Superchargers around the country. Superchargers output higher voltage and thus charge a car at about 400–500 miles per hour – well, kind of. The first 80% of the charge is at that rate, and the last 20% is at one-third of that speed. A third of the way into our Santa Fe trip, we stopped at a Supercharger in Salida, Colorado. We didn't need a full charge to make it to Santa Fe, and 20 minutes later we were on our way.

Our hotel in Santa Fe had a 110-volt outlet in the parking garage that they let us use for free, and thus the car got an additional charge of 40 miles a day while it was parked. (Some hotels offer what Tesla calls destination chargers, which are basically 220-volt outlets equipped with Tesla's small charger, which charges a Model 3 at about 30 to 45 miles per hour – plenty of charge if you are staying overnight.)

Was this as convenient as an ICE car? Of course not. Tesla's Supercharger in Salida was in the middle of nowhere, so we could not go to lunch and ended up walking around for 20 minutes. If we had needed a full charge, it would have taken an hour. For our drive back, we found a Santa Fe Supercharger located in the parking lot of a shopping mall, so the kids and I spent 20 minutes in a coffee shop.

At the beginning of the trip, I was mildly concerned about running out of juice before we reached Santa Fe. However, we were accompanied by perfect weather, and the car guesstimated within just a few miles how much energy it would take for us to reach our hotel. There are about 110,000 gas stations in the U.S., so you can be quite mindless about keeping gasoline in your car. With an electric car, you don't have that luxury – yet.

Side note: Three years ago, on the same drive from Santa Fe to Denver, we discovered that there is a 60-mile stretch that doesn't have a single gas station. We were driving on fumes and ended up buying gasoline from an entrepreneurial restaurant owner at $10 per gallon. (She stored gasoline just for such occasions.) We gladly paid the price, and I explained to the kids that while some people might call this gouging, it was capitalism at its best. If she were driven by altruism, she would run out of her stored gas very quickly and we'd have been stuck in the middle of nowhere, calling for a tow truck.

We need to transport ourselves out of the "gas station" paradigm when we think about EV charging. I'd like you to think about what it takes for oil to be extracted and turned into gasoline. Oil is drilled from deep in the ground in an often desolate place on land or the ocean and transported, by pipeline or boat, often thousands of miles. Then it has to go through a complex purification process (refining) to be converted to a form that gasoline engines can utilize. Next, the refined gasoline has to be transported hundreds more miles. It has to be stored and then trucked again to gas stations, which are basically retail stores that sit on top of very flammable 12,000- to 24,000-gallon gasoline tanks. The fact that we actually have 110,000 such locations around the country amazes me.

When people think about charging electric cars, the first thought that comes to mind is: "So you are going to put charging stations at gas stations. You are going to have long lines of people waiting to charge their cars, since it takes much longer to charge an electric car than to fill a car with gas. It will never work."

We need to change the paradigm.

EVs are plugged into the existing electrical infrastructure that powers everything around us. Electricity is ubiquitous, so you are no longer tied to a station with tanks buried underneath it. Everyone in the U.S. (except the homeless) lives in a dwelling, most of them in houses that have garages and thus their own "gas stations," just as I do. I suspect homeowners with garages will be the early adopters of EVs – this is two-thirds of the country.

But let's say you live in an apartment and you don't have a garage. Most of us spend a few hours a week shopping at grocery stores. Once penetration of electric vehicles increases, you'll have charging stations in grocery parking lots. You'll park your car to go shopping, plug it in, come back 30 minutes later, and your car will be charged.

Capitalism will take care of building out the charging infrastructure. Here is my prediction: At some point, we'll have a charging station mini–stock bubble as companies raise capital and do a land grab. Grocery stores will use charging stations to attract customers, and soon those that can't offer that service will lose customers.

Come to think of it, charging stations will be in all parking lots, from restaurants to office buildings. EV charging will turn into another revenue source for parking lot owners.

I guesstimate that it costs about $8 to fully charge my car at home (about 2.5 cents per mile) and $18 (6 cents per mile) at the Tesla Supercharger. My local utility charges 11 cents per kilowatt, and Tesla charges 24 cents. Thus, Tesla is making a 55% gross margin, or about $10, on every Model 3 customer who "fills up." (Model X and Model S owners charge their cars for free.)

I can see the following business model emerge. Let's say you spend $2,000 per parking stall for equipment and getting the space wired. You make $10 of gross profit per hour. You pay a third to the parking lot owner. You get to keep $7. Let's say each stall is utilized five hours a day (a number I just picked from thin air); that's $12,000 a year per stall in pretax profit, or a 6x return on investment.

Of course, Tesla is charging 24 cents per kilowatt because it wants us to use its Superchargers only when we really need them, so that 55% gross margin is not sustainable and will decline. Today we still have almost 30,000 laundromats in the U.S. – charging EVs is a much better business. EV charging will be the next gold rush, and traditional gas stations will go in the same section of history books as technologies that we have a hard time explaining to our kids – phone booths, cassette tapes, flip phones, horse stables.

The transition from ICE cars to electric vehicles is a bit like the transition our ancestors went through when society switched from horses to ICE cars. At first, people were wondering how they were going to "feed" those cars (it was a lot easier to find grass than gasoline), whether they would have enough decent roads to actually drive anywhere, and whether cars would be crashing into pedestrians and each other. It is obvious to us now that transitioning from horses to cars required a very different type of thinking – a completely new paradigm.

The domain of horses came with an ecosystem that was simply not applicable when we switched to the domain of ICE cars. Even though both performed the same function – horses got people and goods from point A to point B, too – ICE technology was fundamentally different, and so was its ecosystem. I will come back to this very important analogy later.

I imagine the 110,000 gas stations that keep our ICE cars humming along today will look like a rounding error when we count up the millions of electric "filling stations" that will be located in our garages and parking lots.

Part 4: Tesla's Risky Gamble & The Future of Lithium-Ion Batteries

The main character in an ICE car is the engine. Car companies spent billions of dollars to make it do more with less, and they are running into decreasing marginal returns. They have to keep spending more and more to keep getting less and less.

Because the battery (especially Tesla's, as you'll see shortly) is one of the key parts in the EV, it needs to be romanced tenderly to maintain the battery's charge and longevity. Your iPhone is optimized for duration of charge but not for battery longevity. First of all, Apple has an incentive to build planned obsolescence into its iPhone – it wants you to replace it every three years. Second, most iPhones don't spend much time sitting outside in very hot or cold weather; they mostly remain in the comfort of your pocket, at a battery-friendly temperature. And finally, the cost of replacing a battery in your iPhone is $70, but replacing a Tesla battery costs $10,000.

The lithium ion battery – the battery used in your smartphone and Tesla – basically works on a chemical reaction and is sensitive to hot and cold weather. In cold weather, a battery loses its charge faster; hot weather damages a battery, reducing its charging capacity and shortening its life. As the battery is one of the most important assets of the EV, the car constantly micromanages its battery – in hot weather, it cools it down; in cold weather, it warms it up.

Therefore, your EV driving range will vary in different temperatures by as much as 30%. In addition to the car diverting energy from the battery to maintain the battery's temperature (cooling it down or warming it up), energy is also diverted to maintain the temperature in the cabin. Unlike an ICE car that takes heat generated by the engine and turns it into warm air for the cabin, an EV uses its battery to create warm and cool air.

This brings us to what Tesla enthusiasts call phantom drain. When you are not driving and your EV sits idle, the battery charge declines. If you think about it, the same thing happens to your iPhone. Even when you don't use it, the phone is still connected to cell towers and Wi-Fi, and thus the battery still loses power. In the EV, in addition to using energy to arm the security system and maintain a constant connection to a wireless network, the car is protecting the battery when it is hot or cold. I estimate that my Model 3 loses about 12 miles a day to phantom drain, costing me around $110 a year, or the cost of two oil changes.

The engine in an EV, though it will incrementally improve over time, is a less important player. The battery is the most expensive single part of the car and a very complex one, too. Tesla doesn't own the battery cell technology (AA-looking battery units) that goes into its batteries; that belongs to its partner, Japanese conglomerate Panasonic. Tesla designed the battery pack the enclosure that houses the battery cells) and the battery management system controller (computer) that routes and manages electricity flow and the microclimate of the battery cells.

The battery is a key technology for Tesla, but as of right now Panasonic is in control of a big part of it. Just as Apple chose to bring development of the CPU that powers its iPhone in-house, Tesla, which is vertically integrated, may eventually increase its control over its battery technology. The company's purchase of Maxwell Technologies, which has a battery technology that may significantly lower the cost of cell manufacturing, is the first move toward independence from Panasonic.

On the one hand, this strategy has a great appeal because if Tesla is able to produce a better (more durable, lighter, longer-range, faster-charging) battery at a lower cost, it could become a source of a competitive advantage. Today Tesla doesn't fully control its destiny when it comes to batteries, so if BMW decides to use Panasonic's cells, Panasonic will gladly sell to it. BMW would still have to develop its own battery management controller, though.

On the other hand, this vertical-integration strategy could backfire. If EV batteries turn into a commodity and the aforementioned features become ubiquitous, then the lowest-cost manufacturer wins. Tesla would argue that vertical integration will ultimately result in lower costs. The company has built a giant battery factory in Nevada that it calls the Gigafactory. When it is fully operational, the Gigafactory will be able to manufacture twice the quantity of lithium ion batteries produced globally today. Tesla owns the building, and Panasonic owns the cell manufacturing equipment.

Traditional ICE automakers that are tiptoeing into EVs have taken a more conservative strategy and are relying on suppliers (LG Chem, Samsung SDI, and others) to produce a complete battery for them.

(A factoid: Though the battery is called lithium ion, lithium is only 2% of the battery, which consists mostly of nickel and other metals. Rodney Dangerfield would say that nickel gets no respect, and he'd be right.)

I promise this will not turn into a chemistry class, but the battery consists of two electrodes (elements through which electricity flows), the cathode and the anode. Think of the plus and minus signs on your AAA battery. Cathodes carry negative charge and anodes positive charge.

One of the biggest differences between the Tesla battery and the batteries used in other companies' EVs (like the BMW i3, Chevy Volt, and Jaguar i-PACE) is the metals they put in the cathode. Traditional car companies chose the NMC (nickel, manganese, cobalt) combination, while Tesla ended up making a less conservative choice of NCA (nickel, cobalt, aluminum). NCA offers long battery life, quick charging, and great performance. NMC, on the other hand, produces slightly less energy but is less volatile and withstands larger ranges and variations of temperature.

Tesla chose a more potent and more volatile cathode chemistry and elected to control its volatility by trying to manage the macroenvironment of the cells by a special design of the battery enclosure, in order to cool or warm the battery cells as needed. Each battery pack comes with an incredibly sophisticated battery management system that tracks the voltage and temperature of each cell and orchestrates which cells the Model 3 uses.

But let's not sugarcoat this. The Tesla battery has better performance by almost every measure except safety. This is the reason that from time to time you'll see a video of a Model X or a Model S catching fire on social media. In all fairness, though, "Tesla catching fire" makes great headlines because Tesla is the new, exciting kid on the block. If local news reported every time a GM or Ford vehicle caught fire, it would have no time to broadcast other news. According to FEMA, "Approximately one in eight fires responded to by fire departments across the nation is a highway vehicle fire." I suspect FEMA is referring mostly to ICE cars in this statement, as it covers the period 2014 through 2016, when Tesla had very few cars on the road.

A Tesla bull (Tesla's PR department and Elon Musk) and a Tesla bear (a short seller) debated Tesla safety in this article. You decide who is right or wrong.

I choose not to worry about plane crashes when I board a plane. For decades, I chose not to worry about auto accidents and fires when I got into my car that had 20 gallons of combustible liquid in the tank. And today I choose not to worry about an exploding battery in my Model 3. I am doing this not because I am ignorant but because, even if the Tesla bear is right, Tesla fires are still incredibly low-probability events. If I wanted to avoid this along with all the other risks of modern life, I'd be living in a cabin deep in the woods, where I'd still be risking an attack by a not-so-modern bear.

Also, if you take a close look at this short seller's spreadsheet that tracks every death Tesla has been involved in globally since 2013 (including accidents like "Concrete falls on Tesla" and "BMW strikes Tesla"), you'll see that over the past 20 months (2018 and 2019) only four people died in burning Teslas, despite the company's skyrocketing deliveries of Model 3s in that period. It is unclear if they died because of fire or from injuries sustained in the accidents.

Lithium ion batteries are a technology of the late 1980s. They were improved in the '90s and the early part of this century at a somewhat slow pace (especially if you compare them with semiconductors, which have followed Moore's law, doubling in speed every 18 months). The rate of improvement has accelerated over the past decade (in large part thanks to Tesla), and the cost per kilowatt hour (kWh) declined from $446 in 2013 to $127 in 2018. Considering that the Model 3 comes with a 75kWh battery, the approximate cost of the battery has declined from $33,000 to $10,000.

From the perspective of how much it will likely evolve over the next decade or two, EV battery technology is still in its infancy. Until recently, it had little incentive to improve – it was good enough for your flashlight and clock radio. Smartphones and laptops definitely have helped to propel it forward, but their focus is on the duration of the charge more than on the number of charges and dramatically lowering costs (these objectives are a bit mutually exclusive).

As we transition from ICE cars to EVs, the value of the prize will explode as today car industry sales exceed $2 trillion; tens if not hundreds of billions of dollars will be poured into improving the battery. Future EV batteries will have a longer range, last longer, and charge faster. Tesla has already gone through three reformulations of its battery. My $50,000 Model 3 has the latest version, which happens to charge faster than the $90,000 Model X or the $80,000 Model S that Tesla sells today. (And maybe the Model 3 catches fire less – or maybe that's wishful thinking on my part.)

Let me clarify that - while in the short run battery technology is going to be a very important differentiating factor, in the long run the EV battery will likely turn into a commodity and the differentiating factors will be software and self-driving. An EV is a giant computer on wheels, and historically as computer hardware became commoditized, most of the remaining value was in software.

Part 5: Are Electric Cars Good for the Environment?

My wife loves driving the Model 3, not for all the selfish reasons I like to drive it (it is fast and quite the iPad on wheels) but because she feels she helps the environment. Is she right?

Unlike an ICE car, which takes fuel stored in the gas tank, combusts it in the engine, and thus creates kinetic energy, Tesla takes electricity stored in the battery pack and converts it directly into kinetic energy. That's a very clean and quiet process. However, the electricity that magically appears in our electrical outlets is not a gift from Thor, the thunder god; it was generated somewhere and transmitted to us.

As I write this, I am slightly disturbed by how the topic I am about to discuss has been politicized. I am not going to debate global warming here, but let's at least agree that an excess of carbon dioxide (CO_2) and carbon monoxide (CO) is bad for you and me, and for the environment. If you disagree with me, start an ICE car in your garage, roll down the windows, and sit there for about 20 minutes. Actually, please don't, because you'll die. So let's agree that a billion cars emitting CO and CO_2 is not good and that if we emit less CO and CO_2 it is good for air quality.

Roughly two-thirds of the electricity generated in the U.S. is sourced from fossil fuels. The good news is that only half of that comes from coal; the other half comes from natural gas, which produces half as much CO_2 as coal (though it has its own side effects – it leaks methane). Another 20% of U.S. energy comes from nuclear, which produces zero carbon emissions. The remaining 17% comes from "green" sources, such as hydro (7%), wind (6.6%), and solar (1.7%).

So if tomorrow everyone in the U.S. switched to an EV, and our electrical grid was able to handle it, we'd instantly cut our nation's CO_2 emissions by more than a third – that's a good thing.

Aside from all the reasons above (including going from 0 to 60 mph in 4.4 seconds), the reason I am a big fan of electric is that it gives us choices. Gasoline cars run either on oil or on oil. Electric cars open the door for alternative energy sources. That flexibility means oil might stop being the commodity that dictates our geopolitics, and that could mean we'll have fewer wars.

There are alternatives to fossil fuels that have much less impact on the environment – and on you and me. There is nuclear energy, for one. To me, this is a no-brainer. Nuclear power plants have little environmental footprint – they spit out steam. But we have had Chernobyl, Three Mile Island, and Fukushima. It is unclear how many people died in those disasters, because the death toll estimates range from a few dozen who died from direct exposure to thousands who died from cancer caused by radiation.

If cooler minds had prevailed, we would not be on the fourth generation of nuclear reactors but the four hundredth. Nuclear should be our core energy source: It is cheap; it produces very little CO_2; it provides stable, predictable output; and the latest versions are safe (they cool themselves down if there is loss of power). However, what I have learned in investing is that what I think should happen doesn't matter; only what will happen matters. Here is the good news: Nuclear production is expected to increase by almost 50% over the next 20 years, and 90% of the increase will come from the two most populous nations – China and India.

After the Fukushima disaster, Germany decided it wanted to quit nuclear by 2022 (ironically, Japan is going back to nuclear), and its green (wind, solar, hydro, biomass) energy generation went from less than 20% in 2011 to 44% in 2018. Its electricity price to households went up 24% over that period. Germans pay almost three times as much for electricity as Americans, while their CO_2 emissions have not budged.

This happened for two reasons. First, wind turbines and solar replaced nuclear, which produced little CO_2; second, Mother Nature is not predictable, even in Germany. On a cloudy day, solar output drops, and if the wind doesn't blow, wind turbines don't turn, and "peak" coal- or gas-fed power plants have to come online to fill in the gap. Peak plants are smaller, usually less efficient power plants that produce more CO2 per kilowatt hour than larger plants.

Despite the overall jump in Germany's electricity costs, electricity prices in Germany turned negative (yes, negative) on more than 100 occasions in 2017; customers were paid to use electricity. It was breezy and sunny, so renewables produced a lot of energy, but there was no demand for it, so it was cheaper to pay consumers to use power than to disconnect wind turbines and solar panels from the grid.

Ten or 15 years ago, the thinking was that the solution to our energy problems would be the price of solar dropping. It has, falling by 50% to 80% over the past decade, and this trend will likely continue. The problem is that when we need heat or AC in our houses, we expect it to be available instantly, even on cloudy days. Instead of peak plants, we need peak batteries; on very sunny days, we'd store extra electricity in battery packs.

Interestingly, Tesla provided a 100-megawatt battery system to the state of South Australia for $66 million to help it deal with a massive blackout in February 2017. It is unclear if Tesla actually made money on this transaction. However, it was reported that the battery saved the state $17 million in six months by allowing it to sell extra power to the grid.

Recently, Tesla announced the Megapack (a battery pack that looks like a shipping container), a large, utility-grade solution that has 60% greater power density than the battery used in South Australia. Thus, the billions poured into battery technology for EVs are having the unintended consequence of lowering the cost of peak batteries.

There is another inconvenient truth. Batteries and renewables in general are made of earthly materials that have to be mined, transported, refined, and turned into finished products, which are often "browner" (or whatever color is the opposite of green) than nongreen alternatives. The Tesla battery weighs about a thousand pounds and requires 500,000 pounds of raw material to be unearthed to build it.

And there is yet another issue with lithium ion batteries. They require cobalt, a mineral that is found in the Earth's crust. But 50% to 70% (depending on the source) of cobalt reserves are in the Democratic Republic of the Congo (DRC), a politically unstable country that doesn't shy from inhuman labor practices and child labor. Tesla and Panasonic have been reducing the amount of cobalt used in their batteries; it has declined by 60% – the Model 3 battery contains only 2.8% cobalt (the Volkswagen ID.3's battery contains 12% to 14% cobalt). Tesla and Panasonic recently announced that they are working on a cobalt-free battery; they'll substitute silicon for cobalt.

I have to confess that after researching this topic I now understand why Jeff Bezos believes we should mine asteroids for minerals and is committed to spending all his Amazon wealth on Blue Origin, his aerospace venture. And Elon Musk's SpaceX seeks to relocate earthlings to Mars. No matter what path we take in generating electricity, we will still produce plenty of carbon dioxide in the process. However, electric cars provide options (some are better than others), where gasoline cars are making chose between oil or oil.

What Is Going to Happen to Oil?

Today 70% of oil is used for transportation. In other words, we produce about 80 million barrels a day globally. Of that, 60 million barrels go to cars, trucks, and airplanes, and the rest goes for plastics, heating, and other uses. As EVs start taking market share from ICE cars, the demand for oil will decline.

If in five or 10 years EVs are 20% of the total cars on the road, then demand for oil will drop by 12 million barrels a day. Oil production requires a significant up-front investment, and thus oil companies have a strong incentive to pump as much oil out of their wells as they can for as long as they can, as long as their marginal extraction cost is below the price of oil. But supply exceeding demand will eventually lead to a collapse in oil prices.

Low oil prices will lead to bankruptcies, which in turn will lead to decreases in the supply of oil. Oil prices will stabilize temporarily, only to face further declines in demand as EVs continue to take market share. (At some point, national governments will put in ICE phaseout deadlines, which will accelerate EV adoption.) Oil prices will decline again, and the vicious cycle will continue.

Part 6: Why the Survival and Dominance of Car Manufacturers is Far from Certain

What you really should have done in 1905 or so, when you saw what was going to happen with the auto is you should have gone short horses. There were 20 million horses in 1900 and there's about 4 million now. So it's easy to figure out the losers; the loser is the horse. But the winner is the auto overall. 2000 companies (carmakers) just about failed.

– Warren Buffett, speaking to University of Georgia students in 2001

Can traditional automobile companies successfully transition to making EVs?

Today EV sales account for a tiny rounding error of total global car sales. Let's mentally transport ourselves to the late 1800s, when the streets were still busy with horse-drawn carriages and the occasional passing automobile scared a horse or two.

If you cannot relate to a century-old analogy, let's go back to something that happened just a bit more than a decade ago. In June 2008, when the iPhone 3G was introduced, Nokia was still the largest phone maker in the world. What we did not know at the time was that Nokia was actually the largest *dumb* phone maker and that Apple was about to become the largest *smartphone* maker – a small but crucially important nuance. What we did know at that time was that smartphones were the future.

In theory, nobody knows more about making cars than the traditional ICE carmakers – the General Motors of the world – and thus EVs made by

29

these companies should be the ones busying our streets a decade from now. A natural continuity from what we already know may be the easiest cognitive model for us to process, but it is not always the most accurate one.

In 2004, Nokia missed the flip phone boom and lost market share to Motorola, which came out with the slick Razr flip phone. Nokia had a few quarters of disappointing sales, the stock declined, and we bought it. Then Nokia came out with its own flip phone and the status quo was restored: The company was again the king of the dumb (actually, let's be politically correct – mentally disadvantaged) phone castle. The flip phone was a technological change, but it was still in Nokia's domain of core competency. The stock ran up and became fully valued; we made money and sold it. We patted ourselves on the back.

Now, the mistake many investors made, including yours truly, was not seeing that although the iPhone was still called a phone, it was not really a phone but rather a portable computer that, in addition to doing a lot of other smart things, also made phone calls (which the first iPhones were not really good at, but the people who owned them didn't really care). It was not Apple that dethroned Nokia, not at all. Nokia did it to itself. Nokia should have looked at the iPhone and blown Apple a huge air kiss, thanking it for showing the future of "phone," and then gone on to develop its own smartphone.

I made the mistake of applying my 2004 mental model to our Nokia purchase in 2008. With the introduction of the iPhone, Apple took a mentally disadvantaged phone and pushed it into a very different domain with a very different ecosystem.

Assets turn into liabilities

Nokia was a very efficient designer and manufacturer of phones that had very little software and limited functionality. In 2008, the company employed thousands of engineers who knew a lot about wireless signals, plastics, moldings, coatings, and so forth. But collectively they knew little about CPUs, software, and user interfaces. Nokia tried to respond to the iPhone the only way it knew how – by taking its Symbian operating system designed for low-IQ phones and trying to remold it into a smartphone operating system. That attempt failed miserably. We realized what was happening later than we should have and gave up a good chunk of our 2004 Nokia gains.

I never thought I'd say this, but knowledge is not always an asset. *When you are in the middle of a transition from one domain to another, your knowledge of the past domain may cloud your vision.* You'll be seeing through the lenses you're used to wearing.

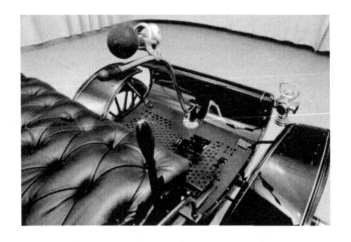

(tiller on early car)

When the first cars were made, they didn't have steering wheels, they had tillers, because they were made by a horse carriage manufacturer. Though it was possible to transition from making horse carriages to making cars, most companies did not; they were stuck in the old "buggy" domain and did not switch to the new "auto" domain.

It is difficult to kill your cash cow

Clayton Christensen discussed this concept in his book *The Innovator's Dilemma*. When your core business is minting money, it is difficult to create another business that may be future-proof but will undermine your core business, especially if the threat is nascent at the time and seems far away. These threats are usually nascent and far away.

When Amazon was practicing e-commerce on books, everyone believed Barnes & Noble would be able to suffocate the tiny company because B&N sold more books in a day than 1997 Amazon sold in months. However, snuffing out Amazon would require Barnes & Noble to lower online and possibly in-store prices, which would hurt its very profitable store business. Well, we all know how that story ended.

You need to have the capacity to suffer

Recode's Jason Del Rey wrote a great article that gave us some terrific insights into Walmart's effort to compete with Amazon. In 2016, Walmart paid $3.3 billion for Jet.com. It seemed like a huge sum considering that there were no other bidders and Jet's forecast for 2016 was $1 billion in revenue and no profits. But investors took this as positive news – finally, Walmart had an e-commerce strategy to compete with Amazon's – and Walmart's stock price went up. Jet.com CEO and founder Marc Lore was put in charge of Walmart's U.S. e-commerce effort.

The average Walmart store stocks about 150,000 items. The Amazon website carries about 12 million items – over 300 million items if you count those sold by the company's Marketplace vendors. In addition to great customer service, Amazon offers great prices, incredible selection, and almost-instant delivery.

If you were an Amazon Prime customer, until recently you had to wait a full two days to get your purchase delivered to you (my heart goes out to you). Not anymore; your suffering is over. Amazon will swallow an $800 million annual loss and start delivering goods to you in one day. Amazon can do this because it is a technological beast that is a byproduct of years and billions of dollars of investment in technology and infrastructure. Today Amazon has 110 distribution centers in the U.S. Its management and shareholders have the capacity to suffer an $800 million loss in the short run, knowing that the strategy will widen the moat around Amazon's business.

Walmart has 20 distribution centers in the U.S. For Walmart to catch up with Amazon, it needs to invest, and invest a lot. Walmart's current infrastructure is designed to support Walmart stores but not to deliver millions of products to customers in one day.

The Recode article details the tensions between Marc Lore and the CEO of Walmart U.S., Greg Foran. Though Walmart's online sales are up 40% (in part thanks to grocery delivery), the company is losing $1 billion on $20 billion of online sales, which is about 5% of Walmart's total sales. The company's board and Greg Foran are not happy about these losses. Walmart is still an incredibly profitable and cash-generative company (with net income of about $7 billion and free cash flows of about $12 billion), but it is a company that has never lost money before. Losses are not in its DNA, while losses are Jeff Bezos' middle name. To make things more complicated, Greg Foran's bonus is tied to profits of both offline and online stores, while Marc Lore's is not.

You can see how difficult it is for even a company as dominant and successful as Walmart to adapt to a shift in domain. In addition to aligning incentives, it needs to have a capacity to suffer (lose money).

Few companies can do that, especially when the existential threat from a new domain seems far away. Walmart stock is owned because of its "safe" and growing dividend. Growing cash flows are safe; declining cash flows and a stagnant dividend are not safe. Walmart's shareholders will revolt if its profit takes a significant fall, even though that may be required to secure the company's future. This point is paramount. Walmart needs to invest billions of dollars and lose billions of dollars for a few years, not because Walmart needs to grow but because if it doesn't invest it will turn into … well, the retail graveyard continues to expand: Kmart, Sears, RadioShack (though not officially dead, they are in hospice care), Circuit City, Montgomery Ward, and the list goes on.

Another important point: Amazon did not build 110 distribution centers overnight sometime in the warm summer of 2018. It built them gradually, year after year. At some point, Amazon had two distribution centers, then it had five. Walmart executives were probably laughing about Amazon's losses; when investors asked if Walmart was concerned about Amazon, they probably answered, "Well, Amazon's sales are X, and ours are 20X." But stupendous sales in a domain that is at risk of melting away may mean very little in the domain that is coming to replace it. The same was true for horse carriage manufacturers, Barnes & Noble, Nokia, and … this brings us to Tesla.

The transition from ICE cars to EVs is not just a technological shift within a domain. It is not like the transition from two-wheel-drive sedans to four-wheel-drive SUVs; *it is a radical shift into a new domain*. I laid out this very extensive domain-shift framework to show that the success of ICE manufacturers in this new domain is anything but guaranteed. Let me expand this framework even further.

ICE cars are low-IQ phones, and Tesla's Model 3 is an iPhone 3G. Cars last about 12 years and phones two to three, so this transition will happen in slow motion.

Part 7: Nokia or Samsung? Why a First-Principles Approach Will Be Key for Developing EVs

As I discussed, during the transition from one domain to another, *many of the assets and much of the knowledge from the old domain become liabilities in the new one.*

Tesla created its cars by entirely breaking out of the domain of existing auto manufacturers. Actually, though this is true for the Model S and the Tesla cars that followed, it was not the case for Tesla's first car, the Roadster. When Tesla first attempted to make an electric car, it was constrained by resources. It wanted to experiment with battery technology and electric engines and did not want to design a complete car. So Tesla adapted the body and powertrain of the Lotus Elise, a sporty gasoline car. Later Elon Musk confessed that had been a mistake – he compared it to keeping the outside walls of a house but gutting and rebuilding the inside, including the foundation. You might as well build a new house.

Because Tesla created the EV industry, it had the advantage of acting from first principles. It could start thinking with a blank piece of paper, not redrawing what already existed. In an interview, Musk said, "I tend to approach things from a physics framework ... physics teaches you to reason from first principles rather than by analogy."

Warren Buffett's version of first principles is "What would Martians do if they landed on our planet?" Not because of Martians' enormous IQs but because they would be new to our planet and could see with clarity things we often don't because we've been here so long.

The first-principles approach allowed Tesla to build EVs that are free of the limitations of gasoline-car thinking. No gears, a skateboard chassis, two engines, a frunk, a credit card key, a mobile app that works as a key and controls the car, and no start button, among others – Tesla applied first-principles thinking to how its cars would be sold. The Model 3 feels like it was designed starting from a completely blank piece of paper and this thinking extended beyond the car and spilled over to selling and servicing the car.

Today's ICE auto manufacturers are basically wholesalers of their cars to auto dealers that are their franchisees. This business model is a Great Depression relic that went basically unchallenged until Tesla came along. The model worked well for automakers and dealers for almost a century, though the experience most consumers had did not fit the definition of well.

Tesla decided that the traditional business model was not appropriate for the new EV domain. Instead, it borrowed a model from Apple, which controls the full customer experience, from buying a phone to servicing it to upgrading to a new one. Also, electric vehicles have fewer parts than ICEs and thus should break a lot less (at least in theory – time will tell), so the traditional dealer model that relies on service revenue doesn't work well for EVs.

This journey of opening its own stores was anything but easy for Tesla. It had to fight opposition by ICE carmakers and local dealers in every state, just as Uber had to fight taxi monopolies.

My purchase of a $51,000 Tesla was as easy as my purchase of a $900 iPhone. I test-drove it. A few days later, I called the Tesla store and told the salesperson that I wanted to buy a car. My information was already in the system; I had to provide it when I placed my deposit in 2015, and I had to confirm it when I scheduled a test drive. I just told the salesperson the configuration I wanted and placed a fully refundable credit card deposit. (I was traveling, but I could have done all this from Tesla's iPhone app or website.) A few days later, I got an email confirming my Model 3 delivery date and asking me to schedule a time to pick up the car. On June 29 at 9:30 a.m., I appeared for my car; by 9:40 I was driving back home. It was that simple.

Tesla changed how a car is serviced, too. A few weeks after I bought the car, its speakerphone stopped working – people could not hear me. I went into the Tesla iPhone app and requested service. I was given a choice between bringing my car to the Tesla service center or having a service technician come out to me. I chose the latter. Two days later, the technician showed up at my office. I gave him my car key and went back to doing research. An hour later, my car was fixed. Tesla's technician had simply restarted my computer. In hindsight, I could have called Tesla and my speakerphone issue could have been fixed remotely.

Now compare these buying and servicing experiences with buying and servicing an ICE car.

It is difficult for ICE companies to adapt first-principles thinking, as it requires them to unlearn what made them successful in the old domain. They are going to have to retool their factories (the smallest challenge of all). They will need to go through a significant and painful change of their workforce. Their current employees have a very different skill set and look at the world through petrochemical lenses (which explains why GM's first foray into electric was the Volt, an electric car with a gasoline engine attached).

Auto dealers, which are an asset to car companies today, will turn into liabilities tomorrow, as Tesla's direct distribution and service model should provide a cost advantage once it gets to scale. Tesla's model is more customer-friendly and efficient, allowing the company to capture the profit that ICE carmakers have to share with their dealers. Because a good chunk of Tesla's cars are built to order, the company doesn't need massive inventory sitting on giant parking lots. Also, ICE manufacturers may not be able to replicate Tesla's direct-sales business model because they are stuck with the franchise agreements they signed with their dealers.

Some ICE automakers are already waking up to the importance of electric vehicles. BMW's CEO announced that he is stepping down. Industry insiders explained his resignation by BMW's loss of market share to Tesla – the Model S and higher-end versions of the Model 3 are direct competitors to BMW cars – and, most important, by the company's lack of progress on the EV front. Recently, BMW announced that it will bring 12 electric cars to the market in 2023, two years earlier than expected.

It won't be easy for ICE carmakers to adapt first-principles thinking to their EVs, but they may not need to: They can copy Tesla. The existing players are not automatically doomed. William Durant, who turned struggling Buick into General Motors, originally made his millions on horse-drawn carriages. Samsung, which was a very large dumb-phone maker, successfully transitioned into one of the biggest smartphone makers today. Samsung did a great job of copying the iPhone, with Google's help; instead of developing its own operating system, Samsung used Google's Android.

Understanding the enormity of the needed investment, carmakers are creating alliances. Ford and Volkswagen are working together on artificial intelligence (AI) and skateboard chassis for EVs. Historically, such alliances in the auto industry have had mixed success.

Traditional car companies have a lot of things going for them. Their strengths are in the designing, assembling, and marketing of cars. They use hundreds of suppliers to make the parts that go into their cars. They can do the same thing when it comes to EVs. They can outsource the battery to LG Chem or Samsung. They can outsource software design to the likes of Cognizant and DXC (we own both). They can use Waymo's self-driving software and Nvidia's self-driving hardware. The traditional automakers are in their best financial shape in decades and thus have capital to finance the EV adventure. Just like Walmart in its new domain, they can afford to make an enormous investment in EVs and take the losses that come with them. But will they? I don't know.

To some degree, their job is more difficult than Tesla's. They have to keep innovating as they make horse carriages – sorry, I mean ICE cars – because ICE cars are what pays their bills. At the same time, they have to focus on the future and invest enormous amounts of time and capital building EVs. When Hernán Cortés invaded Mexico, legend has it, he ordered his army to burn all its boats. He wanted his soldiers to fight as if there was no way back. This is how Tesla is approaching EVs – no boats. ICE companies today seem like tourists in EV-land, with comfortable (ICE) cruise ships waiting for them offshore.

For ICE automakers to succeed in electric vehicles, they need to set up separate EV units with management reporting to the board of directors. The EV management team should be given a blank check, equity in the new company, and the ability to hire people from inside and, most important, outside the company. The existing ICE business should be run with a focus not on growth but on maximizing cash flows. It is easy for me to write this, but it will be very difficult to do, considering that these companies will need to introduce new, exciting cars every four years and entice consumers to buy them, just to keep financing their losses on EVs.

Part 8: Autopilot: Musk's Wishful Thinking or Tesla's Greatest Advantage?

When you buy a Model 3 today, it comes with the Autopilot feature; you can pay $6,000 for Full Self-Driving package. Let's start with Autopilot. Tesla has another, more appropriate name for it – assisted steering. It keeps the car in the proper lane and maintains distance (through automatic braking and acceleration) with the car in front of you. Even when Autopilot is not engaged, it works in the background, nudging the car to stay in its lane and automatically braking if the car detects that you are about to collide with another vehicle.

I have mixed feelings about Autopilot. It can be liberating. On our trip from Denver to Santa Fe, we used Autopilot 80% of the time, as most of the road was a clearly marked highway.

However, when you drive on a road that doesn't have a clearly marked median, Autopilot is not to be relied upon and, in fact, can be very dangerous. On two occasions when the road curved and the lane markings were interrupted by an intersection, Autopilot almost took me into incoming traffic; I had to intervene.

Autopilot is almost ready for prime time. The problem is that the word "almost" should never be used in the same sentence with "Autopilot." An almost-working Autopilot is like an almost-working airplane – it might lead to deaths. This is why self-driving reliability standards are measured four or five digits to the right of the decimal point. Autopilot has to be reliable 99.99999% of the time. Achieving the 9's to the right of the decimal point is much more difficult than to the left of it.

In a few years, Autopilot will be very reliable and saving lives. But we are not there yet. Its presence may give people false confidence in the system and may cost lives. This is why Tesla wants you to keep your hands on the steering wheel even when Autopilot is engaged. If the car detects that you have not had your hands on the wheel for 30 seconds, it will start beeping. If you ignore the warning, the car will slow down and stop.

My advice to you: Listen to the car, not the people who created it. Elon Musk has on many occasions (here is one example) driven the car in media interviews without holding his hands on the wheel, implying that Autopilot is infallible. It is not – at least, not yet.

I use Autopilot only in bright daylight, on clearly marked roads and highways, and I keep my hands on the wheel. I use the hands-off approach only when I am stuck in stop-and-go traffic and moving at very slow speeds.

With full self-driving (FSD), the car should function just like an Uber driver to take you to the destination you enter in the GPS. Well, today's FSD is anything but that. The car will change lanes on its own if you turn on a blinker. It will exit the highway for you if you have put your destination address in the GPS. With the exception of a few other gimmicks that are not yet fully functional, that's about it for now.

Full Self-Driving

The car industry breaks down the levels of self-driving like this: Level 1 (L1) – the car is completely operated by a driver; L2 – the car partially drives itself, but the driver is there as a fail-safe (this is basically Tesla's Autopilot, or assisted steering). The scale tops out at L5, full self-driving, in which responsibility is completely removed from the driver.

Today Tesla and Waymo (Google's autonomous driving unit) are the leaders in self-driving and have taken very different paths. Waymo has several hundred fully autonomous cars roaming the country collecting data so that software and hardware can be improved. So far, Waymo has collected data for 10 million FSD miles, which is an incredible achievement.

Compare this with Tesla's approach. First, unlike Waymo, which uses LIDAR as the main sensor (think of it as radar shooting out light beams),

Tesla uses eight cameras. LIDAR is extremely accurate, but it is bulky and very expensive. Elon Musk is not a big fan of LIDAR; he calls it "a fool's errand." Tesla's cameras are very cheap, and they have long range and high resolution (great for machine learning), but they are not as dependable as LIDAR and don't do well in extreme weather. This is why Tesla is supplementing the cameras with radar and ultrasonic sensors.

Tesla has over 400,000 vehicles on the road today that are collecting data on semiautonomous driving; thus far, it has compiled data for 1 billion miles. Like Waymo, Tesla is using machine learning to train its self-driving system.

I think FSD is a long-term goal and that we are going to have to settle on semiautonomous driving for a long while before FSD gets good enough to be trusted by society and regulators.

Let me tell you a story. I have a friend who is a radiologist. He works from the basement of his house, where he reads X-rays and provides diagnosis eight hours a day. Each day, he reads about 100 X-rays. About one to two of his daily X-rays are randomly selected and reviewed by his peers to catch mistakes. If you think about it, X-rays are a perfect medium to be run through machine learning. You have a very discrete dataset – humans are usually made up of the same organs. If you take a few million X-rays and mark them up (identifying tumors, etc.), machine learning should be able to do a great job of spotting cancer and other anomalies.

My friend, who is only a human, is prone to make mistakes early in the morning before he has had a few cups of coffee or at the very end of the day, when he is tired. Computers don't need coffee, nor do they get tired, so in theory, armed with a proper algorithm, they should make fewer mistakes than humans.

However, our society may not be ready for computers to handle this life-and-death task. At least, not yet. And that is OK, but it doesn't mean our advances in machine learning should completely go to waste. Instead of computers replacing radiologists, they should assist them. First of all, they should do peer review of all X-rays, as the incremental cost of this review is zero. They may catch radiologists' mistakes, and if they come up with false positives, that will just generate more data points that will only make the AI smarter.

Second, computers can assist radiologists in real time by highlighting areas of possible concern. Instead of completely replacing radiologists, they can reduce errors and help radiologists do more. In the process of assisting, they'll get better, and at some point (years and years in the future) they'll be able to replace radiologists.

This is the route Tesla is taking. Despite Elon Musk's promises, FSD and robo-taxis may be quite far in our future. But that doesn't mean Tesla's Autopilot cannot help drivers stay in their lanes and maintain a proper distance from the vehicles in front of them. In the meantime, Tesla is collecting millions of miles of data every single day from its growing fleet of cars. For instance, the Autopilot of my Model 3 will not stop the car at a red light or a stop sign, but it will alert me if I am about to run a red light.

However, Tesla is identifying those stop signs and red lights in what Musk calls shadow mode: Tesla identifies the good drivers among the 400,000 people who drive its cars (I am pretty sure yours truly did not make the cut). Then it compares the decisions the cars would have made with the decisions its good drivers made at intersections. Nobody else can do this kind of analysis today, especially on this scale.

Last year, close to 30,000 people died in auto accidents in the U.S., and three or four died while Tesla's Autopilot mode was engaged. We are used to forgiving human failings, but we don't cut our computers that sort of slack. This gap between humans and computers means that for FSD to be approved by regulators it has to be an order of magnitude better than humans in any weather and any road conditions. Despite Elon Musk's optimism, we are far from that goal.

In the meantime, Tesla's large and growing fleet of cars is constantly collecting data, and with every mile the cars get slightly better. This data and the algorithms that come out of it may turn into Tesla's largest competitive advantage.

Part 9: Elon Musk's Notorious Promises: Why Tesla's Future Success Might Not Be in His Hands

Elon Musk is unlike any leader we've ever seen. For most people, building PayPal would have been an incredible achievement. Not Musk; he built Tesla, a vertically integrated car company and the first new U.S. car company to survive this long since 1930. But Tesla is not just a car company that is on its way to producing a few hundred thousand electric cars a year. It is a company building a charging network that covers not just the U.S. but Western Europe and parts of Asia. It's a company that designed a self-driving processor for its own cars. I can keep going, and I will. Elon also built SpaceX, a company that is able to send rockets into space at a 10x lower cost than NASA – and land the rockets back on a barge parked in the middle of the ocean.

He is working on a company, Neuralink, that wires your brain to a computer. OK, I'll stop. I'd argue that even if Tesla goes bankrupt tomorrow, Elon has succeeded – he has accelerated human progress decades into the future.

Elon Musk (along with Tesla stock) is polarizing in the investing community. He sets impossible goals and achieves half of them several years late. The other half he either will not achieve or just hasn't achieved yet – we don't know. There appear to be two motivations behind his seemingly unrealistic goals. On one side, just like Steve Jobs, he has created a reality distortion field and convinced people with his charisma and confidence that they can achieve seemingly unachievable dreams.

But there is another side to Musk.

Tesla has lost money every year of its existence. (A note from the future: things have changed somewhat since I originally wrote this analysis at the end of 2019, but concepts discussed here still apply. For my up-to-date thoughts, head on over to TeslaAnalysis.Com/Update). There is nothing wrong with this, considering that Musk has created a vertically integrated company in an industry that did not exist 10 years ago. The problem is that these losses have to be financed, and they can be financed by either debt or equity. A company that is losing money has a limited capacity to borrow because it has a limited ability to even pay interest on that debt. Therefore, it has to finance its losses by issuing equity.

Tesla is not trading on a multiple of today's or even tomorrow's earnings; it is trading on a multiple of Elon Musk's dreams. At times, Tesla, which manufactured 245,000 cars in 2018, has had a greater market cap than General Motors, which produces almost 10 million cars a year. The bigger the dreams, the larger the company's market cap and the easier it is for Tesla to finance its losses ... and to achieve Elon Musk's dreams. But distinguishing which dreams will turn into realities is incredibly difficult, not just because of the physical attainability of those dreams but because to achieve them Tesla will need more capital.

So, to project whether Tesla will achieve Musk's big dreams, you need to have an opinion not just on their attainability but on the question of whether others will have faith in Tesla and Musk long enough to allow the company to issue cheap equity to finance those dreams. Every so often, to keep the dream valuation going, Musk makes an announcement that seems to defy what we believe is possible today.

In May 2019, Musk announced that he was expecting a network of Tesla robo-taxis to be operating by 2020. Even if Musk is wrong by two years and this ambitious dream comes true in 2022, it will transform the auto industry and turn Tesla into a cash cow (while also putting Uber and Lyft out of business, unless they come up with their own robo-taxis by then). Or does the robo-taxi belong to a category of sci-fi dream that may or may not be feasible even by 2030?

Before you answer this question, let me give you two more examples.

In July 2017, Musk announced that Tesla was ready to take orders for solar roofs, implying that the technology was ready for prime time. Two years on, after consumers have put down deposits, little more has been heard about solar roofs. Maybe we should have taken Musk literally, not figuratively – Tesla was ready to take orders (but the roof was not ready).

Another example. In his 2016 shareholder letter, Musk wrote that he expected Model 3 production in the fourth quarter to hit 5,000 cars a week (260,000 cars a year). Now imagine you are in the fourth quarter of 2017 and Tesla is producing 186 cars a week (about 9,700 a year). Did Musk lie to you, or did he just dream big but fall short of that dream (so far)? Well, would you believe Musk's next promise, to produce 500,000 Model 3s in 2018 – a 50x increase?

Let's skip 2018 and fast-forward to 2019, where in the second quarter we find Tesla producing at an annualized pace of 288,000 cars and on its way to reach the 500,000 annual production run rate by late 2019 or so.

Given these facts, when Musk makes a forecast, what is plausible and what is exaggeration? How do we know?

These examples illustrate a few points:

1. Some of Musk's promises cross into the unpalatable territory of exaggeration. (Are they really lies if they happen five or 10 years late?)
2. Even promises that Musk will deliver on will come a year or three late.
3. Looking at the past few quarters brings little value when analyzing Tesla. (We have a hard time processing 50x increases in production, even if they come a year later than promised.) Just remind yourself that this is the same person who was able to accomplish things that NASA could not.

To make things more complicated, there are external factors like interest rates, the economy, and trade wars, any of which could cripple Tesla's valuation, thus making the company's financing a lot more expensive or completely undoable. In other words, Tesla is a path-dependent company:

47

Its success will depend in part on factors that are completely outside of its control.

Part 10: Is Tesla Theranos?

Tesla bears – a lot of them are people I respect; some of them are my dear friends – would point out that there is yet another side to Musk, the (fraudulent) Elizabeth Holmes side.

Holmes was CEO of Theranos, a Silicon Valley startup with a board of directors that was beyond reproach, including two former U.S. secretaries of State, Henry Kissinger and George Shultz. At the height of its hype, Theranos had a private market value of $10 billion. Holmes graced the covers of business magazines, gave a TED talk that was watched by millions, was a role model for young women, and had the reputation of a visionary who was going to improve the lives of billions with a product that would be able to run hundreds of medical tests from a small drop of blood.

Holmes may have had great aspirations, but the difficulty or impossibility (at least today) of what she was trying to do caught up with her and she had to resort to outright deception and fraud. Theranos went down in flames.

Bears argue that we are blinded by the Iron Man side of Musk and don't see the Elizabeth Holmes side: that on top of making promises he doesn't intend to keep, he is actually playing a confidence game. Bears point out that Musk committed securities fraud (stock manipulation) when he tweeted that he was taking Tesla private at $420, with "funding secured." This was a lie to scare short sellers out of Tesla stock.

In another such move, Musk personally called a short seller's employer, threatening the employer with a lawsuit if the employer didn't stop the short seller from posting negative research about Tesla on Seeking Alpha and Twitter. (Elizabeth Holmes had resorted to similar tactics.)

Tesla bears also point to Musk's bailout of SolarCity – a heavily indebted, money-losing solar company with a broken business model, run by

Musk's cousin. It would have gone bankrupt in a few months if Tesla shareholders hadn't bought it. A bankrupt SolarCity would have damaged Musk's reputation as the Iron Man who succeeds at everything and thus would have tanked Tesla's valuation and cut off the company's ability to issue cheap equity. By bailing out SolarCity, Musk heaped a money-burning operation and billions of dollars of debt onto Tesla, which was already struggling to break even.

The bears questioned Musk's state of mind (sanity) when he called a British diver rescuing Thai kids trapped in a cave a "pedo" in a tweet.

I want to remind you that I started out this essay quoting F. Scott Fitzgerald and explaining that in our analysis of Tesla and Elon Musk we needed to maintain "two opposed ideas in mind at the same time and still retain the ability to function."

There are a lot of opposed ideas here.

Skeptics have plenty of substance in their arguments against Tesla. History is on their side. Other than Tesla, the last car company to be started in the U.S. and survive to the present day was born in 1930, and two car companies went bankrupt during the 2008 financial crisis. Yes, Musk's stock manipulation with the "funding secured" tweet was simply immoral and illegal. We can vent about it, but the outrage people feel over that behavior (yours truly included) is irrelevant to Tesla's future. Musk paid a fine to the SEC and is still running the company. (Any other CEO probably would have been fired and might have gone to jail.)

However, the issue raises a more important question: Is Musk an asset or a liability to Tesla? Well, he may be both. On one side, he is a genius and a visionary who deeply cares about Tesla's success and is willing to sleep in the factory to get things done. But he also is an incredible micromanager and a benevolent dictator who runs four companies. He is overworked and exhausted, and this to some degree explains (though it doesn't excuse) his erratic behavior and his tweets, too.

If something were to happen to him or he was removed from Tesla, what would happen to the company? Would Tesla turn into Apple after it fired Steve Jobs in 1985 or the Apple of 2011 after Steve Jobs passed away? The Apple of 1985 withered because it did not have enough breadth of products or the depth of management needed to replace a charismatic visionary. Jobs left a vacuum of leadership and vision when he was fired, and John Sculley – a Pepsi executive – was not up to the task.

The Apple of 2011 was in a much stronger position and much more broadly based, with a stronger leadership team. Tim Cook was handpicked by Steve Jobs and groomed for years to replace him. And though Apple as a business has done well since Jobs' passing, the company's subsequent innovation has been very evolutionary, not revolutionary. It has mostly improved the categories of products developed under Jobs but has not come up with a significant new product category (aside from the Apple Watch). In fact, it has tried and so far failed at producing a car.

The Apple example shows that replacing a benevolent dictator/visionary is very difficult but not impossible. Success depends on timing, the company's competitive and financial strength at the time, depth of management, and luck. Tesla today seems to lack depth of management – it's shedding executives faster than I am losing my hair. Today Tesla is closer to the Apple of 1985 than that of 2011 and can hardly afford to lose its overworked genius/benevolent dictator.

Tesla's success will be determined by the financial viability of its business: Can the business finance itself? Tesla's success as a stock will depend in the end on the company's earnings power.

The bears argue that Tesla has quality and service issues. Some Model 3s that were rushed out to meet production goals were poorly put together. It seems that the Model 3 production quality issues have been resolved. Even one of Tesla's biggest critics, Bob Lutz, ex-CEO of GM, has raved about the quality of the Model 3's design and assembly.

Tesla is growing at the speed of a Silicon Valley startup, but in addition to writing a lot of software, it is building Gigafactories, a global Supercharger network, service centers, and stores, and designing its own self-driving processor – all while competing against companies that are better capitalized and on the 50th iteration of their ICE products (and thus have more consistent quality).

Every time Tesla has stumbled, and it has stumbled plenty, it has gotten up, regrouped, and moved forward. My personal experience with Tesla's service has been excellent. But I read that today Tesla is still experiencing growing pains with its service. This makes sense – its serviceable car fleet has more than doubled over the past year and a half. The Tesla technician who fixed my Model 3 speakerphone told me that his scheduling software still needs work, as he is sent to back-to-back appointments that are 50 miles apart. This doesn't sound like a permanent issue, though; a quick software fix should resolve it.

One very important thing that Tesla has and Theranos did not is an incredible product and people who are fanatical about it. Tesla cars are superior to other EVs and actually to most non-EV alternatives made by the competition today. Talk to any Tesla owner and he (it is usually a he) will rave for hours about how much he loves the car. I have yet to meet a GM or Toyota owner who has the fanaticism of a Tesla owner.

This is why Tesla spends zero on advertising (while ICE carmakers spend billions) but has a bigger sales force than all GM, Ford, and Chrysler dealers combined – Tesla has passionate owners, and its sales force is growing with every car sold.

Today it is really hard to tell whether Tesla will reach escape velocity and turn profitable before investors and capital markets lose their patience and willingness to finance its losses. By buying SolarCity, Tesla certainly made its journey more difficult. But the company is not sitting still and is trying to cut costs. Here is one example. The Model S had two miles of electrical wire, the Model 3 has "only" one mile, and the Model Y is expected (if it can actually be implemented) to have only 100 meters.

If Tesla is able to increase its production, the incremental cost per car should decline. Let me explain. The cost of a car has two components, fixed and variable. Variable costs include the battery, tires, engine, etc.

These costs don't usually decrease much (though they do decrease some) as a larger number of cars are produced.

Fixed costs, like running a Gigafactory and developing software, decline on a per car basis as Tesla increases its production. Assuming demand for Tesla's cars continues to increase, the company's gross margin (how much it makes per car) should increase, and thus it will reach profitability.

However, even in the worst case, if Tesla runs out of money and bond and equity investors lose faith in the company, it is unlikely to follow the fate of failed car companies (think DeLorean or Tucker). It will be bought by an ICE automaker or a Silicon Valley comrade (Google or Apple). It would save them billions in losses on R&D and getting up to scale. Investors would still lose money, of course, because at that point the distressed purchase price would be a fraction of today's price.

Part 11: Is Tesla Apple? Is Elon Musk Steve Jobs?

Throughout my analysis, I have compared the Model 3 to the iPhone, Tesla to Apple, and Elon Musk to Steve Jobs. What I am struggling with is this: Can Tesla become as successful as Apple, and can Tesla cars turn into an iPhone-like franchise, taking EV market share from nothing to 10% to 30% of the ICE car market?

Tesla has many advantages. It is not burdened by assets from the previous domain (gasoline cars). It is incredibly focused. Going back to our Cortés analogy, it has no boats waiting for it. If this electric thing doesn't work out, it is done; there is no plan B. Its vertical integration may work to its advantage, as it is competing against companies that have to rely on their suppliers and their messy alliances.

Tesla's first-principles approach will allow it to be constantly years ahead of the competition. Despite its zero advertising budget – its advertising department is an army of one, Elon Musk, with 27 million Twitter followers – Tesla has one of the most iconic global brands. It is run by a relentless founder who is willing to put in hundred-hour weeks and sleep in the factory when needed. Tesla is years ahead of the competition on battery development (I am not 100% certain about its exclusivity, though) and software. And then there is self-driving (with 1 billion miles of data), which may provide Tesla a lead that will be difficult for its ICE brethren and even Google to catch.

The company's competitors, despite their strengths, also have the weaknesses of being profitable, dividend-paying companies that therefore have a lower tolerance for sustained losses (see the Walmart and Nokia discussion in Part 7).

But is Tesla another Apple? During the first few years of its existence, the iPhone was far superior to any other phone on the market. Today smartphones running on Android are often cheaper and may have better features than iPhones. However, through software differentiation (the iPhone runs its own operating system) and the creation of an ecosystem (iMessage, Facetime, and Apple Music, which run only on iPhones and other Apple-made devices, and which work incredibly well with one another), Apple was able to create frictional switching costs that keep iPhone users upgrading every two or three years to new iPhones.

Like other automakers, Tesla doesn't have switching costs or an ecosystem that locks you in. Yes, there is brand loyalty for some, but most car buyers are one commercial or one test drive away from switching to a competitor's product. Will this behavior change in the new domain of EVs? The incredible feeling that you get when you drive an electric car, the jolt you get from instant torque, is not going to be unique to Tesla. The answer to the question we're asking lies to a large degree not with Tesla but with ICE carmakers' ability to transition to the new domain of EVs.

Tesla bears would argue that ICE automakers will be coming out with dozens of EVs over the next few years. So far, reviews I've read on EVs made by traditional carmakers have been favorable, but the software interface still looks and feels like Nokia trying to patch its dumb-phone Symbian operating system onto a smartphone.

Just as the iPhone was initially not competing much against other smartphones but was really up against dumb phones, Tesla is not competing with the new EVs but with the 86 million ICE cars sold globally today. If Tesla survives in the short run and turns profitable, then I can see a path by which the company could start producing a few million cars a year.

I can tell you one thing: Just as you cannot go back to a dumb phone once you get inoculated by a smartphone, I will not buy a gasoline car ever again. I drove my wife's gasoline car (my pre-Tesla car) a few days ago, and I was shocked by how slow and unresponsive it was. But will my next EV be a Tesla? Today I think so, but tomorrow? Not sure.

Now let's figure out how much Tesla is worth.

Will Tesla's semi-truck or full-self-driving cars see the light of day over the next five years? I don't know, but at that point it won't matter. Let's say Tesla sells its average car for $40,000. Today the average car in the U.S. sells for $38,000, but with the Tesla there are gas savings and little or no maintenance required, so I am giving Tesla a $2,000 benefit of the doubt. In fact, today the average Tesla sells for $50,000.

Let's assume that Tesla will sell 2 million cars a year (half in the U.S. and the other half in the rest of the world). GM and Toyota sell a bit less than 10 million cars each, so Tesla will capture 5% of the U.S. market and a bit more than 1% of the rest of the world market.

If Tesla makes $8,000 per car, that's a 20% gross margin. Today it is at 18%, though this number is subsidized by environmental tax credits. Toyota and GM are at 18% or so. Musk's goal is a 25% gross margin. (He is counting on Tesla's ability to sell full self-driving, which has 100% gross margins and today creates costs but little revenue.)

Now let's say all of Tesla's other expenses – R&D, general expenses, interest, and anything else – are $8 billion a year (today they are less than $7 billion). With the above assumptions, Tesla would have a pretax profit of $8 billion. At 10x pretax profit, its market capitalization would be $80 billion, double today's $40 billion. (A note from the future: while Tesla's stock price has changed significantly since I originally wrote this analysis at the end of 2019, the concepts discussed here still apply. For my up-to-date thoughts, head on over to TeslaAnalysis.Com/Update)

Neither bears nor bulls will like the preceding math excursion. Bears would probably argue that Tesla may not be able to get to 2 million cars a year (because it will be sunk by SolarCity's debt and losses), and if it does, it will not be able to achieve 20% margins. Bulls will argue that the above assumptions are too conservative – if Tesla can do 2 million cars, it can do 5 million.

And in the meantime, while the bulls and bears are arguing, I am fascinated by how my thinking about Tesla and ICE carmakers has changed since I spent a few hundred hours analyzing the industry. I don't know what probability to put on this, but I can see how Tesla can succeed if it reaches escape velocity and starts generating cash flows.

Tesla bull or bear? My up-to-date thoughts

Tesla is constantly changing. New models, new factories, new loans...

So am I a Tesla bull or bear?

What do I think of Tesla's current valuation?

Get my up-to-date thoughts on Tesla stock at TeslaAnalysis.com/update (entirely free).

Or find my latest podcast episode on investor.fm

Made in the USA
Monee, IL
29 October 2020